Management Division 2015

AA003078

Core Programming Area at the 2015 AIChE Annual Meeting

Salt Lake City, Utah, USA
8-13 November 2015

ISBN: 978-1-5108-1864-4

Printed from e-media with permission by:

Curran Associates, Inc.
57 Morehouse Lane
Red Hook, NY 12571

Some format issues inherent in the e-media version may also appear in this print version.

Printed by Curran Associates, Inc. (2016)

For permission requests, please contact AIChE
at the address below.

AIChE
120 Wall Street, FL 23
New York, NY 10005-4020

Phone: (800) 242-4363
Fax: (203) 775-5177

www.aiche.org

Additional copies of this publication are available from:

Curran Associates, Inc.
57 Morehouse Lane
Red Hook, NY 12571 USA
Phone: 845-758-0400
Fax: 845-758-2634
Email: curran@proceedings.com
Web: www.proceedings.com

TABLE OF CONTENTS

430823 Career Planning: Would I Fit or Be a Misfit? Understanding Organizational Culture, a workshop for Students and Young Professionals

Abstract:

This workshop began as an idea presented to the Management Division Leadership Team with the aim fill voids in preparing students to enter the work place, and subsequently was expanded to include young professionals. In general, academic curriculum or career services resources did not offer a way to match one's personal values with an organization's culture. Results of a survey of students and YPs identified the value of a highly interactive workshop which would increase the participant's capability to pick an organization which fits them best.

The Workshop has three basic parts:

- Establish key personal values which must be realized in a job.
- Explore applications of a tool or framework to assess the culture of an organization
- Compare Personal Values to an Organization Culture

In each part of the 2 hour session, concepts are introduced followed by structured exercises for small groups of participants to test the ideas, develop useful applications and personal action items.

The final design of the workshop is the result of the work of a 7 member diverse Project Team, interviews with student leaders, review with YPs, and a live "preview" with students. The evaluations were used to make appropriate improvements to the initial design.

436255 Poster Presentation Success: How to Prepare and Present a Winning Poster

Abstract:

Preparing a poster presentation provides a great opportunity to connect with potential collaborators and employers. The key to success is optimizing your time before, during and after the session, and really taking advantage of the eyeballs on your poster to build a strong network and develop job leads. In this workshop, we'll discuss the elements of a successful poster and how to deliver a meaningful presentation. You will learn how to craft your value statement and talking points, and how to effectively engage those who stop by. You'll also discover how to utilize social media to build buzz around you and your research endeavors, and how to follow up on the connections you make. Speaker: Alaina G. Levine, President, Quantum Success Solutions, Author: Networking for Nerds (2015 Wiley)

405005 Innovation from Beginning to End: Generating Ideas, Working with People, and Managing Projects

Abstract:

Innovation from Beginning to End: Generating Ideas, Working with People, and Managing Projects

Jack Hipple

Principal

Innovation-TRIZ

Tampa, FL

Eldon Larsen

Professor of Engineering

Marshall University

South Charleston, WV

Extended Abstract

This workshop focuses upon managing R&D projects from beginning to end. Beginning refers to mental picture of an idea whereas end refers to solving the customers need and in the process generating profit for the organization. Using a case study, the speakers will delineate challenges involved at every stage of R&D project from a mental image of an idea to a profit making product. The talk also compares old paradigm of innovation management with the new proposed paradigm. The probability of success can be significantly improved using the new paradigm.

Successful Innovation Management

We have seen many idea management flow sheets, gate management processes, and prioritization recommendations starting with an idea that already exists, but no discussion of where the idea came from or what was necessary to generate and frame it. Moreover attention is not paid to shaping and propagation of the idea. This has been especially true over the past decade when processes such as Six Sigma came into vogue whose main thrust was to eliminate deviation. New concepts and innovation are, by their very nature, deviant! In order to generate fresh and innovative ideas we must create a new thought pattern that may challenge the existing status quo in any number of ways. These can even include, at an extreme, replacing one's current products and businesses with totally new approaches and models.

The key elements of successful commercial innovation include:

1. 1. Continuous flow of unmet needs, both from current customers and related to core competencies of the organization ("outside in" as well as "inside out"
2. 2. Using clear mental models of problems and potential solutions
3. 3. Perseverance for idea propagation
4. 4. Clear definition of the business opportunity
5. 5. An appropriate product development and management team
6. 6. Efficient process scale up and continued input from the market
7. 7. Successful project management to insure implementation on schedule and within budget

We will look at each one and discuss the important issues.

1. Continuous Flow of Unmet Needs and Exploration of Uses for Core Competencies

An organization needs to have a process for how it decides what to work on. This may sound simplistic, when this is left to intuition or accident; it is unlikely that the "best" projects will be chosen.

Steps to make sure that an innovation effort is capturing the best opportunities:

- Talk to groups that are not your current customers but that might be able to use the functional capability that you have as a core competence. Customer competitors are another valuable source. The customers of your customers are another excellent source. This insures that you have awareness several levels above your product and customers and alerts you to potential long term needs and trends that may not be obvious to your direct customer or may not be readily disclosed by them.

Figure 1: Level Diagram

In these discussions we want to make sure that we have a clear definition of what the "ideal result" that our customers are desiring. We can also look at parallel universes of technology and markets to explore opportunities.

2. Using Clear Mental Models of Problems and Solutions

It is one thing to "invent" a new product, but another to make sure that we understand how a potential customer is going to use the product as well as how the customer's customer is going to use their product. The days are gone when we can just make something, put it in a drum or tank car, and wait for the invoice to be paid. We must be intimately familiar with the product's use, its limitations (including regulatory issues if appropriate). We must understand the details of the physical use of the product and how it affects a customer's products and business.

It is also important to make sure that we do not have blinders on when thinking about use of our product. It is rare that there is only one way to accomplish a function. We must be constantly vigilant about competitive activities, not just from direct competitors, but from parallel approaches to meeting a customer's needs.

3. The People Side

Ideas, innovation, and projects never happen in a vacuum. They are accomplished with teams of people, all of which have different personalities and styles. Recognizing and using these pro-actively is critical to a project's success. How these are measure and used will be discussed.

4. The Business Opportunity

A new product is not a business unless we have a way of sustaining the product's sales, expanding its utility, and developing and enforcing an intellectual property portfolio. As we commercialize a new product, we must be constantly aware of, and be willing to address with resources and commitment, issues such as change in product form, packaging and shipping issues, product purity, and shelf life/stability. These issues are most adequately addressed by constant and continuous communication with the customer.

5. Product Development Teams

We frequently choose product team members based on "who's available". While this may be expedient, it may not serve the best interests of the new business development. The obvious component of team selection is function—marketing R&D, manufacturing, legal, sales, etc. but we frequently ignore the human side of team functioning. If team members spend too much time dealing with personal issues relating to other team members, that energy is not applied to successful new product launch. Tools that can assist in this include measurements such as the Myers Briggs family of instruments, the Hermann Brain Dominance instrument (HBDI), and the Kirton KAI assessment. Knowledge of these personality profiles will not change natural behavior, but can make team members aware of others' preferred styles and open discussion can eliminate hidden barriers to effective team functioning.

6. Efficient process scale up and continued input from the market

Often times, companies strive for launching the best product. It is important to realize that continued input from the market is critical in order to have that 'best product' in the market. It is necessary to efficiently scale up and launch a product sooner than later. It should be perfected using market feedback. Getting market feedback is the only way to close the R&D loop and serve your customers to the fullest.

7. Successful Project Implementation and Management

New product development does not end when the first semi-commercial sample has been OK'd by the customer. A new type of team is now required to oversee the details process design, plant construction (including all the sub issues of permitting, EPA regulations, etc.). Quick and efficient plant start up with appropriate contingency planning is also a part of this effort. Contingency planning, not only related to unforeseen plant construction and start up difficulties, but also to taking care of the customer's needs in these situations must be part of the plan.

Three keys to successful project management lead to successful project completion. These keys involve

- · The importance of people
- · The importance of defining and setting good objectives
- · The importance of good planning

We will apply these three keys to successful project management. This is all about application, not just theory

435977 How to Entrepreneur - Getting Started

Abstract:

Getting Started

The ah ha moment is the entrepreneurial idea. Many people have many ideas. What separates an entrepreneurial idea from just the muse running though the normal thoughts of the day is the impact the idea may create. Can the idea change the world ? If it can, then is there a way to capture and execute the idea into a sustainable effort. Almost all execution is to deliver a product or service. They both have common characteristics to achieve success. Even world changing ideas need a resource stream. Whether a philanthropically funded or revenue generating idea, the resource flows must be balanced similar to a common mass balance in a production system. The most important part of getting started is just that. In the initial throws of generating support for the idea, you must be able to pitch the idea constantly to various audiences. Many companies have been launched from the famous elevator speech describing the idea. Many others go through the "walk the street" pitch deck. One famous guidance is the 10,20,30 rule developed by a famous entrepreneur. The pitch deck will evolve as the does the idea. Business plans are usually the next step after people latch onto the idea. The issue is to remember developing the product and service is more important to starting than the pitch deck or business plan. The reality is that if the idea is world changing enough, the pitch deck and business plan becomes very simple as it would be embodied in the obvious understanding of the plan by demonstration of the idea.

This session will present the "How-To's" on

1. The Entrepreneurial Idea

2. The Initial Pitch Deck

3. The Business Plan

436061 How to Entrepreneur - Legal Matters, Intellectual Property Basics

Abstract:

Legal Matters - Intellectual Property Basics

Having a great idea that can evolve into a business is a fragile thing until there is a critical mass in cash flow to sustain the business growth. Capturing the idea requires trial error and documentation. The recent changes to the USPTO procedures makes the need for documenting and filing protection for the many ideas that become the kernel for a business even more imperative. The ability to defend claims, to insure that the idea is both unique and novel is guided by a set of legal work flows that allows the creation of a patent. Patents teach everyone what you are doing and gives you the license to protect what you are doing for a limited period of time. Alternative to a patent, is the protection of a trade secret which as long as it remains protected, is a competitive advantage that may be the critical element to insuring that business is able to grow without having competition knock the wind out of the sails of the new business too quickly. All successful business relationships are built on trust. Years of work to develop the subtleties of the relationship then are codified in clear and effective contracts. Every transaction and supplier relationship is a contract. The same base principles are woven into the variety of contracts that are required to establish and operate a business. All transactions are contracts. Good fences make good neighbors, and good contracts make good business.

a. Patents

b. Trade Secrets

c. The USPTO

436002 How to Entrepreneur - Financial Matters

Abstract:

Funding Metrics

What is the math behind all the financial metrics ? The complexity if engineering math is rooted in the physical modeling that presents sense to things unseen and behaviors of the natural world. The math of financial pursuits also present clarity and comparison basis for things unseen and concepts based upon time and the change of opinion. Having a few of the basic definitions helps clarify the financial confusion that can exist due to many different experiences and attitudes being part of the equation. The public perception of the rewards of entrepreneurial efforts can be staggering. The reality can be quite different. Understanding the mechanics of the various ways the rewards are presented and delivered is an important part of the package that is required to make sound personal decisions in your commitment to the various aspects of the entrepreneurial world. Serial entrepreneurs will admit that it takes a few rounds in the tank to fully understand all the ways the rewards can be presented and calculated. Seemingly simple math can become quite burdened with complexity when determining the valuation of the company, the details of stock options, and the various instruments of debt and equity that are presented.

1. EBITA
2. Debt v Equity
3. Dilution
4. Angels, Venture Capital, Private Equity, IPO, Banks
5. A Round, B Round, C Round
6. Valuations
7. Rewards

436059 How to Entrepreneur - Building the Team

Abstract:

Building the Team

High performance team members have specific roles and complementary talents. The team is committed to a common purpose, and works with a high level of collaboration. They are innovative, tight-knit, and built on trust. The team has the collective confidence to surmount any barrier. Leadership is more orchestration bringing out and blending the subtleties of the common purpose of the team. As with all organizational structures, there is “the” leader shouldering the ultimate responsibility, however, everyone on the team is “a” leader. “Leadership” is rotated according to the need at that moment in time. Decisions are made on the analysis of the facts. Egos are checked at the door and there is efficient conflict resolution if it is ever needed. High performing teams could take over the world if that was the common goal to which everyone was committed.

The team will make or break the effort. The level of trust amongst the team members is the best gauge as to whether or not the effort will be a success. The 90% of start up failures can probably trace the death blow back to some team issue. Small is better, and check the ego at the door. When do you exceed your capacity to continue to contribute to the enterprise? Are you a builder or an organizer? Businesses go through different cycles of need. The leadership team that can launch the business might not have the skills to sustain and execute the growth and operationalization phase. Recognizing when you have contributed you best to the phase of the project needed for success and then realizing how to exit with the rewards is critical to closing out and moving on the next exciting opportunity. The myriad of electronic connection tools that are available give an effective, efficient and enjoyable ways to keep the efforts moving without having to use the 1960’s management style of “face” time. Chemical projects will require physical operations. The lab and pilot systems will need to be staffed. These locations could become he rally points for various physically needed for the continued development of the system. Spend the money saved on talent and physical chemical equipment needed.

1. Recruiting
2. Seasoned Leaders
3. Entry Level Passion
4. Interns Payroll
5. Benefits
6. Employment Taxes
7. Virtual Organizations
8. Teams and Structures

436056 How to Entrepreneur - a Young Professional's Adventure in an Entrepreneurial Organization

Abstract:

The leap from the classroom to the field is an exciting one. The thought that finals and late nights in the lab are finally at end breeds eagerness, but this is far from the truth. Entering an entrepreneurial organization has the tests and late nights but in a very different way. One begins to learn to implement the skills accumulated in college. High school taught you how to learn, college is where you learned to think, and the professional world is where you implement. The plant, lab, and jobsite become the classrooms, the operators, other engineers, and office coworkers become the teachers and if you hope to succeed you take advantage of all the resources available. An entrepreneurial job is sink or swim and swimming requires hard work and long hours, if you're willing to do the work and put in the time the reward is unmatched. An entrepreneurial organization is in constant change, continually improving, and sometimes making a few mistakes. A young professional with a positive energetic attitude cannot find a better place to be early in their career. This talk breaks down some important lessons learned in the first year as a young chemical engineering professional in this entrepreneurial environment.

436077 How to Entrepreneur - the Contracts of Doing Business

Abstract:

From day one as an entrepreneur, you will be dealing with attorneys. The paperwork that establishes your business, and other agreements that you will make with investors, suppliers, and clients that operate the business on a day to day basis are all contracts that need some sort of legal advice to ensure that the documents reflect the relationships properly. Some key portions of these contracts, such as stock offerings, board rules and powers, and other investor-entrepreneur agreements are vital to the success of any start-up company. Understanding what you have, and what you may be negotiating (giving) away, are often the difference between you or your investors making that move to a Caribbean island of your choosing. Another type of contract that often becomes the kernel for many entrepreneurial efforts is the intellectual property contract. Which type of intellectual property protection you choose, how you document your ideas, and the agreements between you, your investors, and your engineering staff may make the path from inception of the business to sustainability less hazardous. Patent and trade secret protection, and the relative merits of each, are important marketing and business generating strategies that every entrepreneur should have in their arsenal.

1. Contracts

2. Terms and Conditions

3. Employment Contracts

4. Invention Disclosures

5. Purchase Orders

435979 How to Entepreneur - Measuring the Effort

Abstract:

Measuring the Effort

An entrepreneurial effort requires different measurements for evaluating progress for many different audiences. It is said to expect what you inspect. Measurements are the easiest way to describe progress against goals. No matter what the audience, measurements must be real, valid, and actionable. The engineers designing the system have to be able to roll up the many measurements in the system to how the system affects the financial performance that the investors will be most interested in measuring. Many tools are used, and the resulting trend and data analysis is only as good as the measurements being taken. One of the most mis-understood aspects of a new system is the validity of a measurement. Following good statistical methods will insure that decisions are made on good information. The critical element to the technical piece of the investment is the data that supports the claims. Can the product be demonstrated to be reproducible on the systems that are proposed? Does the development of the data, both technical and commercial, follow sound and accepted practices? The capital investment for chemical projects are large and models must be validated against actual data from bench and pilot systems. Experimental Orders (EOs) and Key Learning Reports (KLRs) are a simple and organized way to define the runs that are required to produce the data that supports the claims being made for the investment. Capital and Operating Costs must be justified to the financial investors to deliver the returns expected. The information is only as good as the measurements used to distill the understanding of the effort. Technical data must be correlated to the financial and commercial returns expected.

1. MEB – Mass and Energy Balances
2. DOE – Design of Experiments
3. Measurement Methods and the Statistics of Variability
4. Data Packages

    1. Experimental Orders
    2. Key Learning Reports
    3. FLRCs (Factors, Levels, Responses, Constants)
    4. Data to Support the Claims
    5. Technical Inputs to Financial Models
    6. Capital Cost Estimate
    7. Operating Cost Estimate
    8. Financial Data Packages

436060 How to Entrepreneur - Legal Matters, Intellectual Property Development in the Academic World

Abstract:

Legal Matters - Intellectual Property in the Academic World

Publish or perish. Generate excitement for the idea that is emerging from the lab. Present to colleagues at credible events. These are all realities of the academic expectations of innovation development. How do you balance the need for disclosure and the ability to protect the kernel of the idea that will become the intellectual property for the foundation of a new business ? The balance is tricky as the rules and procedures for the USPTO work flows may conflict with the need to disclose and show the world progress against academic commitments.

1. Academic IP Development
2. Academic Technology Transfer
3. Licensing

424427 Engineered Systems for Entrepreneurs in Academia and Industry

Abstract:

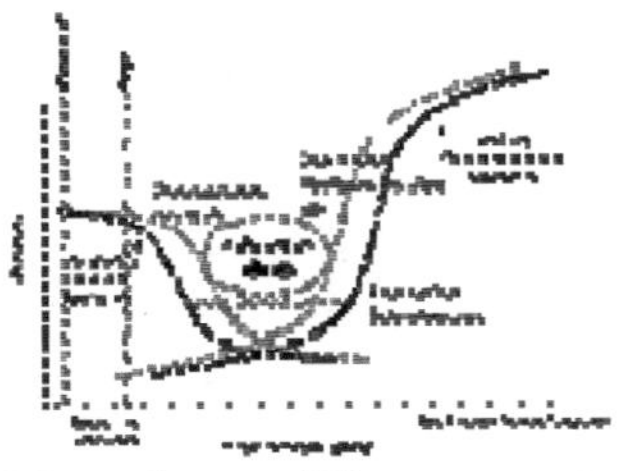

Fig. 1. Discovery to innovation: crossing the abyss.

The Engineering Research Centers (ERC) Program has fostered entrepreneurial translation of scientific and engineering discovery to innovation since its establishment in 1985. The National Science Foundation initiated its flagship ERC program at the request of the White House and National Academy of Engineering to develop an interdisciplinary culture of innovation by partnering academia and industry. Innovation ecosystems cultivated in sixty-one multi-institutional ERC partnerships between universities, industries, and federal agencies in the past 30 years have bridged the 'valley of death' between discovery and marketing (see Fig. 1). ERCs have been awarded 382 patents, issued 669 licenses, and spun out over 171 companies with more than 1500 total employees.

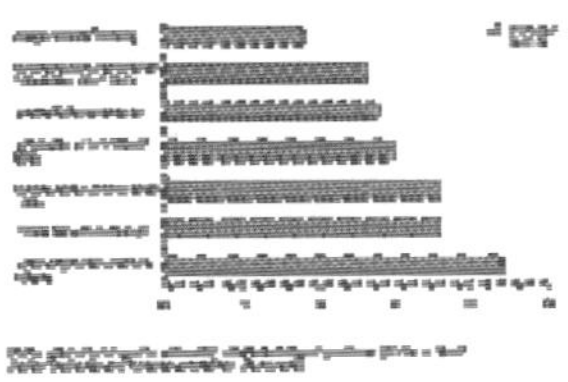

Fig. 2. Graduates of ERC program are better prepared for industrial innovation.

A number of innovative best practices have emerged from the ERC program that provide comprehensive support for entrepreneurship. Among these best practices are novel curricula, interdisciplinary team-based project development, comprehensive strategic planning, vertical and horizontal mentoring by academic and industrial collaborators, experiential learning activities, and rigorous formative and summative assessment. As a result, students who graduate from ERC program are ranked significantly higher than their peers in a number of relevant categories (see Fig. 2). Included among over twelve thousand participants in ERCs in the past thirty years are founders of Google.

Fig. 3. T-shaped education for innovation: discipline, system, and boundary crossing.

Among federally supported programs that foster innovation, the ERC program is unique in its focus on translation of discovery to marketable innovation within the context of an engineered system. Test-beds in each ERC guide foundational research and integration of technology into proof-of-principle demonstrations of an engineered system. Test-bed evolution is accelerated by benchmarking against state of the art, roadmapping using quantitative metrics and milestones, input from industrial and scientific advisory boards, annual evaluation by site visitors, and regular assessment of strengths, weaknesses, opportunities, and threats. Faculty, students, and industry researchers who partner in ERC activities engage in 'T-shaped' acquisition and development of innovative capacity (see Fig. 3).

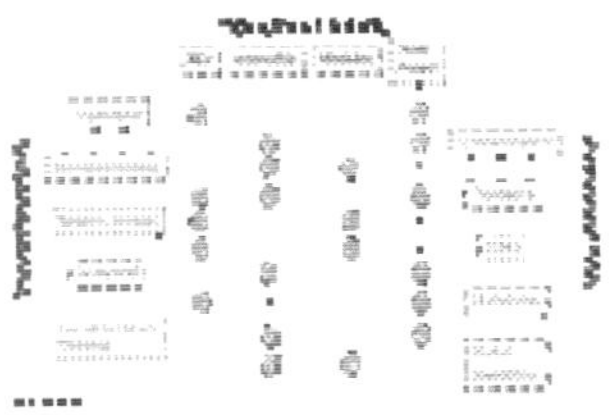

Fig. 4. A roadmap for entrepreneurial development.

Products and processes commercialized from ERC-developed technology are estimated at >$50B, a ~50-fold return on a roughly ~$1B federal investment in ERCs. The Food and Drug Administration recently licensed the Argus II retinal prosthesis which was developed by collaborating academic and industry entrepreneurs in an ERC at University of Southern California. ERC programs and participants are encouraged to develop a personalized roadmap for entrepreneurial development that adapts to the unique needs, challenges, and opportunities in their area of science and engineering, technology sector, and market segment (see Fig. 4.)

This presentation will review the structure and design principles of the ERC program and experiences of its participants that highlight key aspects of entrepreneurship at the intersection of academia, industry, and government.

436071 Mitigating Safety and Regulatory Risk in a Chemical Project Effort

Abstract:

Safety and regulatory risk are inherent to the development of new chemical technologies and processes. Before a product can be brought to market or a process brought online, hazard assessments may be necessary to evaluate the safety of the product or process during intended and unintended use. Performing a proper hazard analysis for a new system is complicated by the lack of available process data. Issues such as corrosion or side products formed during cleaning can be missed when testing at small scales. When developing consumer products, the interactions between the uninformed consumer and a potentially energetic, toxic, or reactive product can also be difficult to predict with small sample sizes and consideration of consumer misuse may be necessary for some products. This talk will discuss the role of code reviews and process hazard analyses in product or process development and discuss relevant case studies.

408542 Applied Fundamentals of Project Management—an Overview

Abstract:

Project Management consists of many principles, skills, and tools that are essential for a chemical engineer to be successful in today's project-oriented workplace. This presentation will highlight the most essential principles, tools, and techniques. Covered will be the three most important keys to successful project management, how to more effectively communicate and work with people, and how to develop effective project objectives and prioritize them. Additionally, a brief overview of important planning tools will be provided. This presentation is highly application oriented, and will be useful to all potential and practicing project managers.

419056 The Application of Balanced Scorecards in Paper Manufacture

Abstract:

The Application of Balanced Scorecards in Paper manufacture

AIChE Annual meeting. Salt Lake City. Nov. 8-13, 2015

Virgilio L. Gonzalez, PhD

E-mail: gbconsult75@gmail.com

ABSTRACT

The Balanced Scorecard (BSC) was created more than fifteen years ago by economists Robert Kaplan and David Norton, and is finding applications in industrial practice. This management tool consists of a chart showing four perspectives with related indicators, which are Financial, Customer, Internal processes and Learning and Innovation. The application of this scorecard is shown in some case studies within the Paper industry in three South American developing countries.

The proposed BSC related to Paper manufacture is shown below.

BALANCED SCORECARD – Paper Manufacture

Perspective	Strategic Objectives	Indicators
Financial	• Capital utilization • Cost reduction • Sales increase	• ROI, NPW, DCFR • % Cost raw materials • % Machinery renovation
Customer	• Satisfaction • Profitability • Retention	• % Satisfaction/production type •% Sales/customer •% Faithful customers
Internal Process	• Innovation • Operational process • Aftersales service	• % Improvement operational logistics • % Plant stoppages • % Product devolutions
Innovating and learning	• Employment Capacity • Emp. Competitiveness • Emp. Satisfaction • Info Systems capacity	• Productivity • % Training • % Employee Satisfaction • Competence level per Employee

Different from bulk chemicals manufacture, where there may be low profitability, from a Financial point of view, Cost savings seem to be as important as sales increases. Additionally, Paper companies should consider capital utilization as a driving force because of ecological pressure and customer needs.

Customer perspective seems to be more relevant in Special papers making because of fashion changes and the need to make differentiated products to compete with Far East manufactured goods. Paper buyers seem to be more of a captive market, although marketing activities have to be pursued. E.g. special quality papers.

Effective internal processes need to be considered in all company categories but can be more critical in Paper industries because they depend on punctual demands, e.g change of fashion or eventual need of special poducts, school activities. Some paper industries depend on agricultural production, although the tendency is to use recycle paper in most cases.

Finally, Learning and Innovation appear to be more relevant in the making of most products innovating to be more friendly to the environment than in technology itself, particularly in the use of raw materials and waste products.

The systematic application of Balanced Scorecards can contribute to better management and to make reasonable decisions about what to do next to keep business alive in spite of current limitations, mainly where there are foreign currency exchange control regulations.

In conclusion, the knowledge and development of Balanced Scorecards can definitely help project managers and directors to have a global vision of a company through its basic perspectives, motivating that projects in their different stages keep alive and operationally efficient.

408539 Effectively Building a Project Management Work Breakdown Structure—How to Do It!

Abstract:

The Work Breakdown Structure (WBS) is onc of the most fundamental and important documents in project management. Yet, strangely, many project managers either do not know about them, skip them, incorrectly construct them, or incorrectly use them. This presentation will teach the fundamentals of how to build and use a WBS for effective project management.

422381 A Systematic Approach to Resource Management

Abstract:

This paper will present an approach for the systematic evaluation of alternatives. The process begins with members of a panel independently and privately ranking the alternatives as well as providing a measure of their own familiarity with each alternative. Various statistical and clustering techniques are used to analyze the data and to generate a ranked list of the alternatives reflecting the collective assessment. Statistical metrics are provided that help determine consistency in the evaluations as well as detect any biases and other anomalies in individual panelist assessments or in the objects under evaluation. This methodology can be applied to project selection as well to performance evaluation. It offers the advantage that it removes skewness that might result from a highly dominant player influencing the group and is thus a more streamlined and more objective process.

An Introduction to:

犹他-青海联盟 Utah-Qinghai EcoPartnership

US-China Sub-National EcoPartnerships

犹他-青海联盟 Utah-Qinghai EcoPartnership

犹他-青海中美绿色合作伙伴

MISSION

TO DEVELOP RELATIONSHIPS OF TRUST THROUGH JOINT COOPERATION IN AREAS OF COMMON INTEREST

犹他-青海联盟 Utah-Qinghai EcoPartnership

犹他-青海中美绿色合作伙伴

MISSION

JOINT RESEARCH, PROTECTION AND TRANSFER OF INTELLECTUAL PROPERTY

犹他-青海联盟 Utah-Qinghai EcoPartnership
犹他-青海中美绿色合作伙伴

MISSION

IMPROVING HUMAN HEALTH AND WELFARE, WITH A FOCUS ON THE ENVIRONMENT, FOOD PRODUCTION AND CLEAN ENERGY

犹他-青海联盟 Utah-Qinghai EcoPartnership
犹他-青海中美绿色合作伙伴

MISSION

TO FACILITATE THE DEVELOPMENT AND IMPLEMENTATION OF MARKETABLE TECHNOLOGIES FOR ECONOMIC DEVELOPMENT

犹他-青海联盟 Utah-Qinghai EcoPartnership

BACKGROUND

Utah-Qinghai EcoPartnership - 2011

ECOPARTNERSHIP signing ceremony in the Treaty Room at the U.S. Department of State with then secretary of State Hillary Clinton. U.S.-China sub-national partnerships provide cooperation in both the private and public sectors on environmentally important issues.

犹他-青海联盟 Utah-Qinghai EcoPartnership

BACKGROUND

Utah-Qinghai EcoPartnership - 2011

Inspired by a common vision, then Governor of Qinghai Province, Luo Huining, and Utah Governor Gary Herbert, developed a partnership aimed at joint innovation for environmental sustainability.

Utah-Qinghai EcoPartnership - Overview

Regular Conferences (annual or bi-annual)

Discuss common needs, make technology presentations and progress reports, plan joint initiatives
(Attendees: Government groups, universities, technology providers, potential funders, UQEP)

Technology Implementation and Commercialization

IP management, investment, new or existing companies, local economic development

UQEP

Not for profit organization sponsored by the U.S. State Department and the Chinese National development and Reform Commission

Technology Identification and Joint Research

Universities, start-ups, private firms

Initiate Demonstration Projects

Determine objectives, set time-table, secure funding, publish findings

犹他-青海联盟 **Utah-Qinghai EcoPartnership**

ACCOMPLISHMENTS

Renewable Energy Project and Bio-waste Remediation

This anaerobic digestion facility represents the successful transfer of technology to efficiently process bio-waste pollutants, converting them into renewable energy in the form of natural gas while reducing greenhouse gas emissions and waterway

犹他-青海联盟 Utah-Qinghai EcoPartnership

ACCOMPLISHMENTS

Soil Remediation Technology Demonstration Project

Technology from the University of Utah was used to clean organic chemicals from the soil of a contaminated industrial site. The soil remediation process called Heightened Ozonation Treatment (HOT), was used in a demonstration project to clean up a vacated industrial site on the shores Wuxi City's Lake Taihu.

犹他-青海联盟 Utah-Qinghai EcoPartnership

ACCOMPLISHMENTS

Heavy Metals Detection Technology

New Technology Research from the University of Utah is behind new mercury detection technology which will help to manage heavy metals in Qinghai and elsewhere.

犹他-青海联盟 Utah-Qinghai EcoPartnership

ACCOMPLISHMENTS

Sand Filtration Process to Clean Drinking W

Completed the Purdue-Utah village water project in two small villages located nearly two hours from the provincial capital, Xining. Purdue University scientists, supported by the Purdue-led U.S.-China Eco-Partnership for Environmental Sustainability implemented slow sand filter technology.

犹他-青海联盟 Utah-Qinghai EcoPartnership

ACCOMPLISHMENTS

Utah-Qinghai-Purdue Joint Water Resource Lab (in planning)

Joint effort between Purdue University and Qinghai Normal University led by the Utah-Qinghai EcoPartnership for ongoing efforts to improve water availability and quality for villagers who rely on drinking water directly from rainwater collection.

犹他-青海联盟 Utah-Qinghai EcoPartnership

犹他-青海中美绿色合作伙伴

ACCOMPLISHMENTS

Utah-Qinghai Joint High Altitude Medical Lab

Conceived in cooperation with the University of Utah and Qinghai University in 2013, completed in 2015. The 65,000 square foot, $5M facility has been sanctioned by the Chinese Ministry of Science and Technology. The lab will contribute to research and cooperation on high altitude medicine between the two

犹他-青海联盟 Utah-Qinghai EcoPartnership

犹他-青海中美绿色合作伙伴

NEXT STEPS

CONTINUE THE UQEP'S MISSION AS THE

UTAH SUSTAINABILITY COMMISSION

- PROMOTE COOPERATION AND TRUST WITHIN GOVERNMENT AND BUSINESS TO ADDRESSES COMMON INTERESTS AND ENVIRONMENTAL ISSUES
- SPONSOR INTERNATIONAL EXCHANGES AND CONFERENCES
- PROMOTE TECHNOLOGY DEVELOPMENT, PROTECTION, APPLICATION AND COMMERCIALIZATION
- SPONSOR TECHNOLOGY DEMONSTRATIONS PROJECTS

416159 Patent Protection Across Borders: Tips for Obtaining International IP

Abstract:

Patent protection across borders requires special attention from inventors and companies and special efforts from legal counsel. Many businesses have lost significant opportunities by failure to properly consider the challenges of international IP protection. We will discuss best practices and guidelines that can help you and your employer be more successful and cost effective in protecting your business in multiple nations.

421131 Innovation with Asian Teams: Overcoming Stereotypes and Cultural Gaps

Abstract:

Innovation with Asian teams can be extremely successful, but often poses problems for Western companies that lack an understanding of the cultural and language barriers that can get in the way. Managing innovation across borders with Asian partners can be most effective when Western managers or team members have been properly prepared. Here we share experiences from our innovation work with Asian teams with an emphasis on recommendations for effective innovation in China and India.

421158 International Innovation, Licensing, and R&D: Panel Discussion

Abstract:

In this panel discussion, speakers will share their extensive international, IP, and innovation experience to provide guidance on how to best manage international opportunities in business, R&D, and IP. Topics covered include how to protect trade secrets and IP (including licensing), how to deal with regulatory burdens, how to identify and overcome cultural barriers, and how recognize when there is trouble hindering innovation.

435929 The Art of Being an Entrepreneur

Abstract:

Entrepreneurs have an exciting approach life. Chaos is welcomed as long as opportunities can be kicked out of the mix and an organized defined clarity can lead to a profitable investment. Risk is not avoided, but is well understood and embraced. The simplicity and focus of delivering the initial entrepreneurial product is followed by well thought out plan for growth. Entrepreneurs are full of paradox. The drive and passion to launch the idea into the business may have to be replaced with letting go to continue its path to growth and continued success. George Washington laid down the sword once he made sure that his start up was on a successful path. The focus on the metrics of investment must be balanced with the need to change direction on new information. Willingness to go all in and know that in any situation a solution will emerge continually battles with the need for insuring that bills are paid and commitments met. Captain Kirk's ability to give the order to save the day, must be preceded by Spock's logical evaluation of the options for consideration. Chemical system development must never compromise on safety. The evaluation and understanding of the energy stored and potential energy released from various operations must be given the time and resources needed to insure that a hap hazard and corner cutting first prototype is not dangerous to all who are involved. This consideration must not overwhelm the other traits that make the chemical development effort entrepreneurial. The spirit must be active and passionate, and it is even more important to have the highest performing people who know what they are doing.

436445 Incubation, Collaboration and Creating the Innovative Environment

Abstract:

Many ideas just need a push. The initialization of a great effort needs a supporting environment to enable forward progress. Entrepreneurs have tough skin. The adventure of creating something new pushes through the multiple rejections and the all and naysayers. Reality is that all ideas need an environment in which to build the idea. The garage is the most popular place for idea incubation. Many great companies have started that way. Creating an innovative environment by bringing together likeminded groups trying to change the world creates and energy. Ideas breed ideas. Support for situational commonality refreshes and recharges individuals to continue the quest. Creating innovative space where many “garage” efforts can incubate creates an acceleration of the additive power of the energy. Incubators and accelerators can be found in the most interesting locations.

436571 Building the Entrepreneurial Adventure

Abstract:

Many successful entrepreneurs evolve into the investment and support side of the business. Developing the street smarts from having been in the middle of rising to the challenges of getting something started becomes some of the best experiential training possible. The discussion and presentation of the how to do it sets the stage for going out and developing the skills to then learn from mistakes and enjoy the adventure that ensues.

One of the greatest gifts a successful entrepreneur can give is the guidance and sharing of their knowledge to those who follow with the same passion and ambition. It is hard work. There are many details that have to be worked to understand. Tell me how to fish and I hear, show me how to fish and I see, let me fish and I then understand. The successful entrepreneur leverages their success as the guidance to others is given and the energy that they create is rewarded a hundred fold in the positive impact that the effort makes on many lives.

An Entrepreneur's Personal Journey

Ricardo B Levy
Lecturer, Department of Chemical Engineering
Stanford University

2015 AIChE National Meeting
The Entrepreneurial Chemical Engineer
November 10, 2015

The business eco-system

Extended Business Model

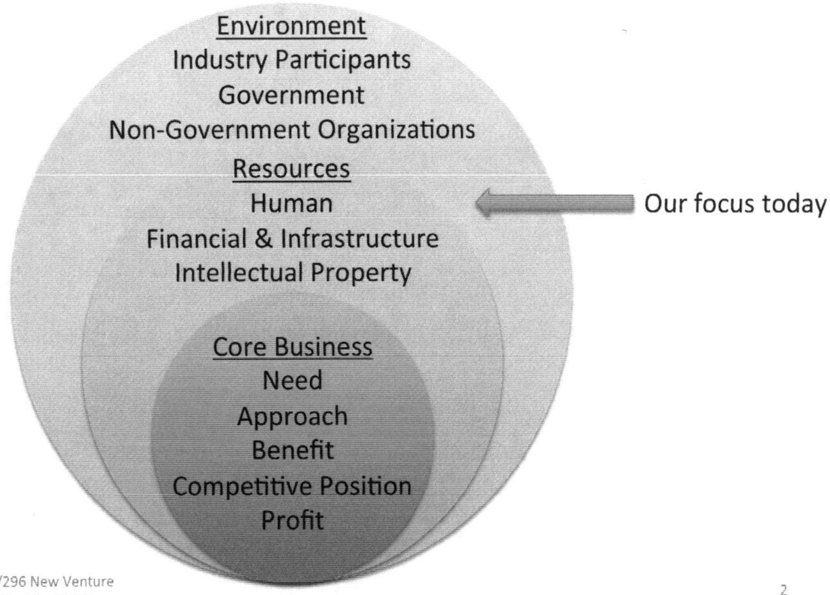

The whole entrepreneur

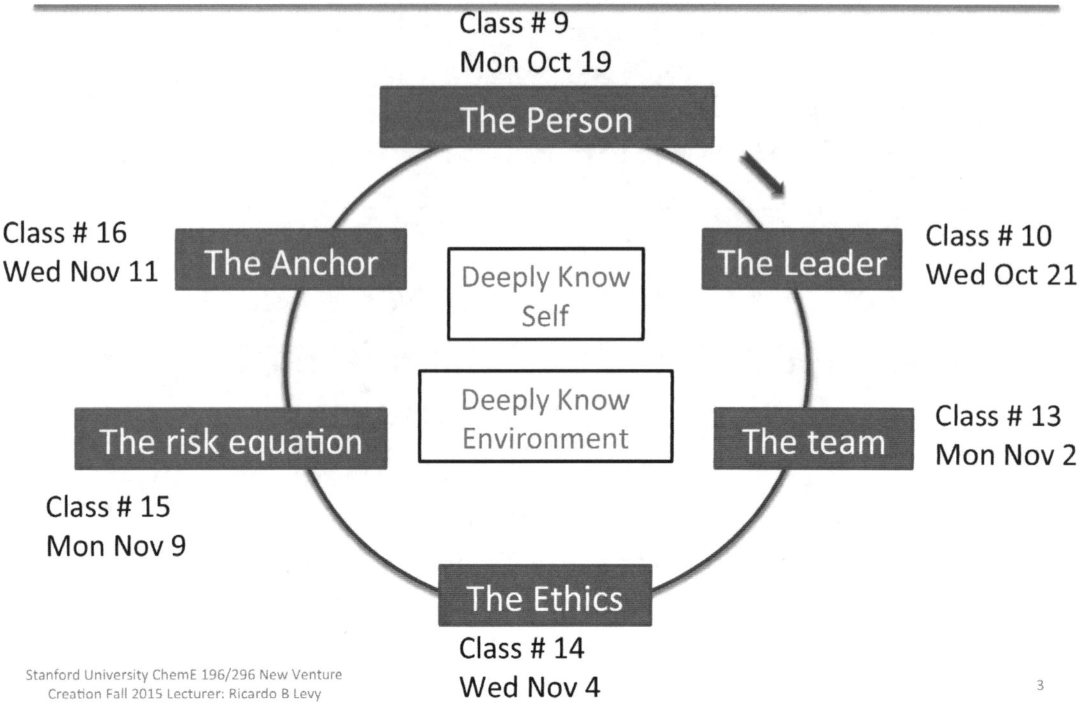

Stanford University ChemE 196/296 New Venture Creation Fall 2015 Lecturer: Ricardo B Levy

3

The Essentials Of An Entrepreneur

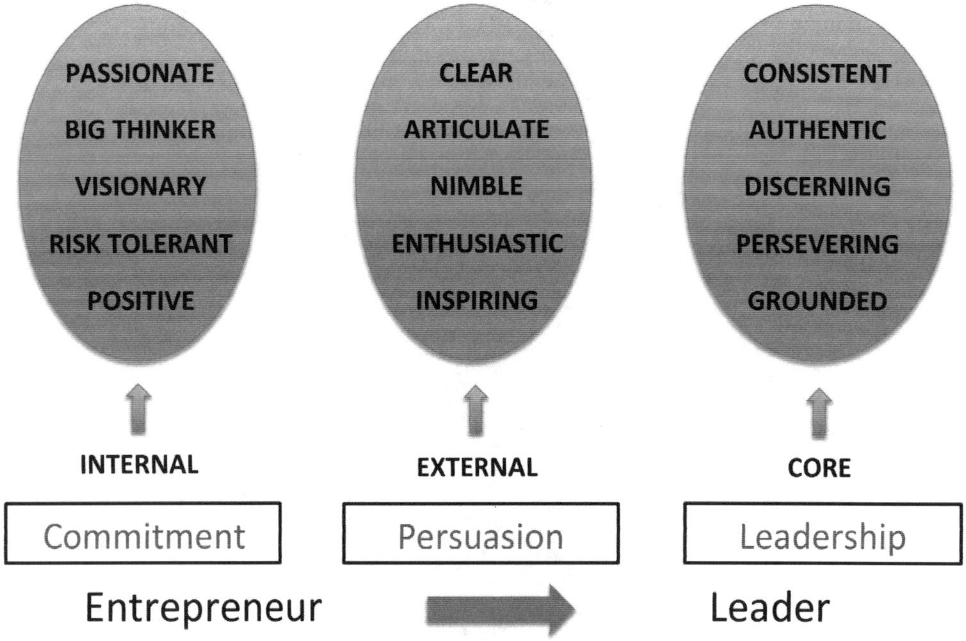

Stanford University ChemE 196/296 New Venture Creation Fall 2015 Lecturer: Ricardo B Levy

4

Internal characteristics

PASSIONATE
BIG THINKER
VISIONARY
RISK TOLERANT
POSITIVE

Must Haves		Give us
Passion	→	Intensity
Purpose	→	Direction
Positive Mind-set	→	Attitude

Commitment Capacity

External characteristics

CLEAR
ARTICULATE
NIMBLE
ENTHUSIASTIC
INSPIRING

Must Haves

→ Story Teller
→ Dancer
→ Lover

Persuasion Ability

Core characteristics

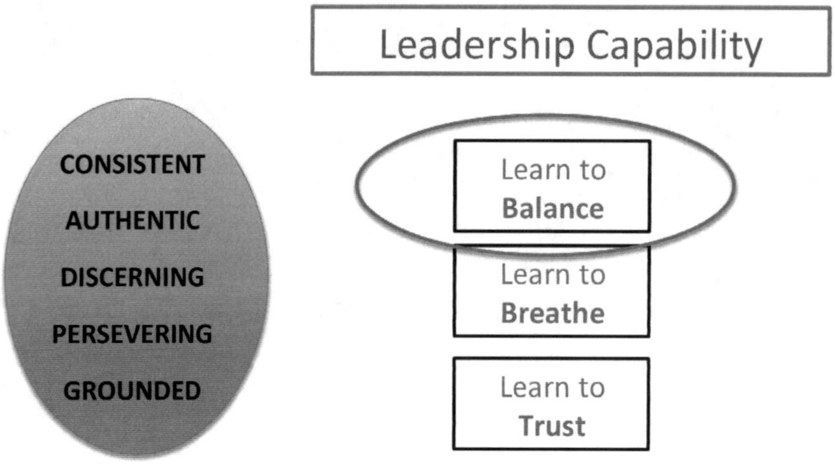

Many Balance challenges

- Optimism vs realism
- Leading vs Managing
- Strategy vs tactics
- Focus vs breadth
- Control vs delegation

Balance: leading vs managing[1]

LEADER	MANAGER
• POINTS	• PLANS
• ALIGNS	• ORGANIZES
• INSPIRES	• PROBLEM-SOLVES
• PRODUCES CHANGE	• PRODUCES PREDICTABILITY

Very different strengths

Entrepreneurs need to do both

1. John P. Kotter, What Leaders Really Do, HBR December 2001

Your "core" characteristics

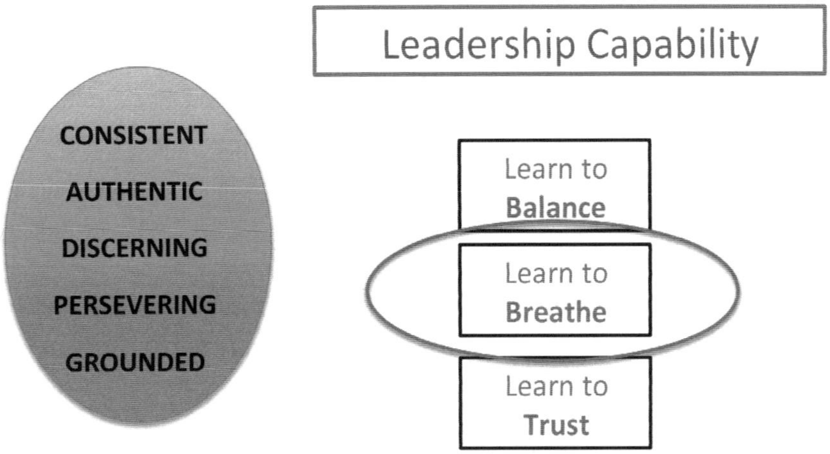

Breathing fundamentals....

When we breathe in

We expand
We encompass
We absorb

Our Horizon grows

When we breathe out

We push
We direct
We focus

Our Attention sharpens

Stanford University ChemE 196/296 New Venture Creation Fall 2015 Lecturer: Ricardo B Levy 11

The buck stops here

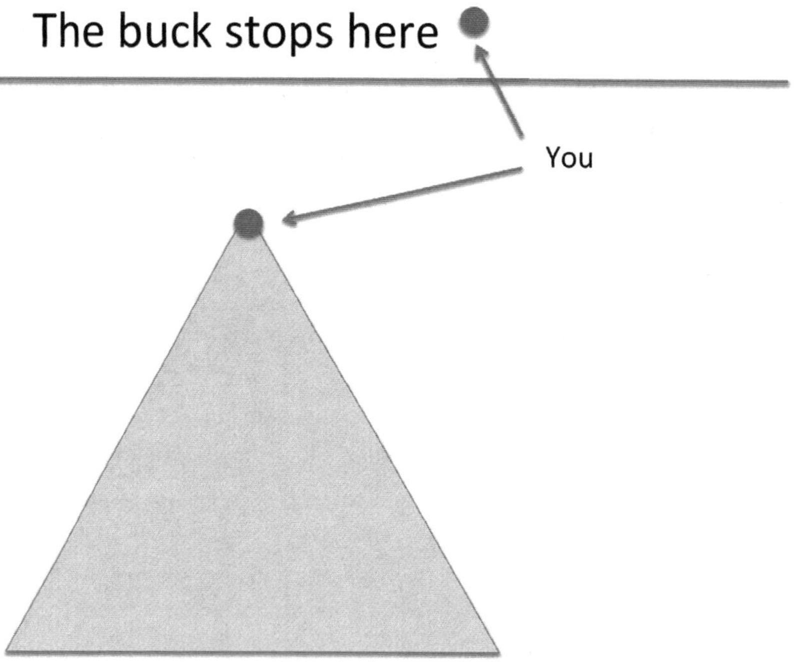

Stanford University ChemE 196/296 New Venture Creation Fall 2015 Lecturer: Ricardo B Levy 12

We have to breathe in

To HOLD the whole picture

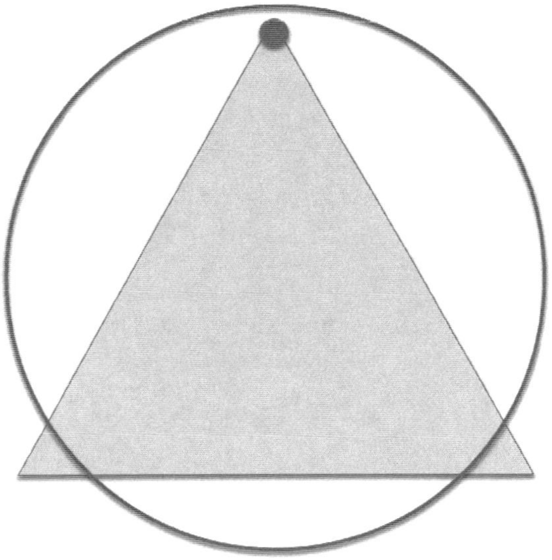

Stanford University ChemE 196/296 New Venture Creation Fall 2015 Lecturer: Ricardo B Levy

13

We have to breathe out

To point our team

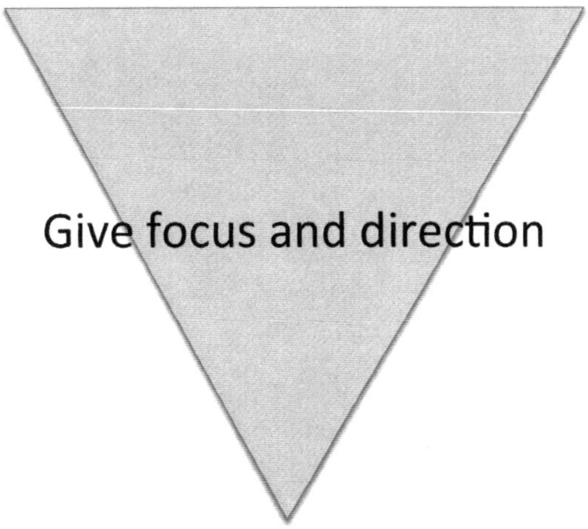

Stanford University ChemE 196/296 New Venture Creation Fall 2015 Lecturer: Ricardo B Levy

14

Breathe in...

We HOLD the whole picture

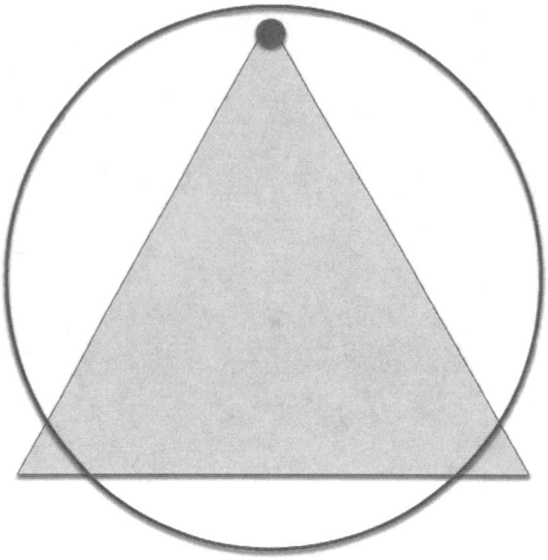

Stanford University ChemE 196/296 New Venture Creation Fall 2015 Lecturer: Ricardo B Levy

15

Breathe out...

We point our team

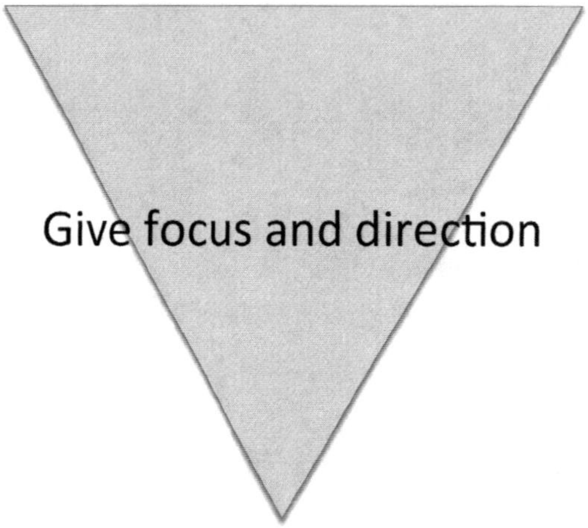

Stanford University ChemE 196/296 New Venture Creation Fall 2015 Lecturer: Ricardo B Levy

16

When we breathe in

We HOLD the whole picture

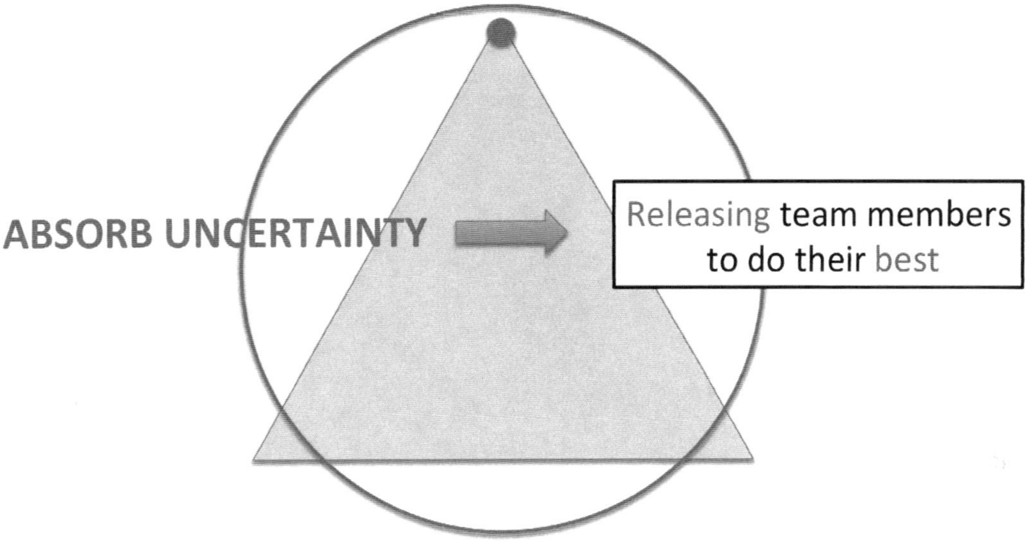

Stanford University ChemE 196/296 New Venture Creation Fall 2015 Lecturer: Ricardo B Levy

17

Your "core" characteristics

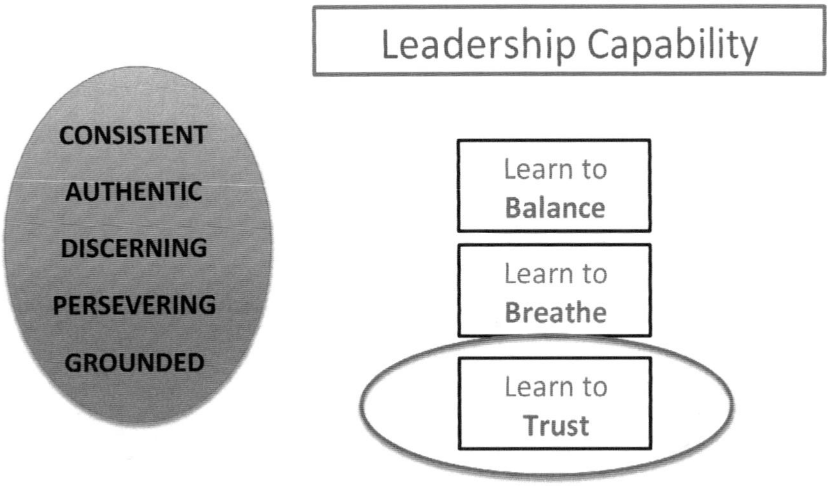

Stanford University ChemE 196/296 New Venture Creation Fall 2015 Lecturer: Ricardo B Levy

18

Absorbing Uncertainty...is all about trust

The fabric of trusts

Beware: it gets heavy...

Stanford University ChemE 196/296 New Venture Creation Fall 2015 Lecturer: Ricardo B Levy

21

... and lonely ...

Stanford University ChemE 196/296 New Venture Creation Fall 2015 Lecturer: Ricardo B Levy

22

How do we cope with heaviness and loneliness...

- Companions in the journey
- Meditation and other centering methods
- Increasing our self awareness
- Avoiding hubris -- staying humble in the face of success

Nourish your relationships

Nourish yourself

Nourish your relationships

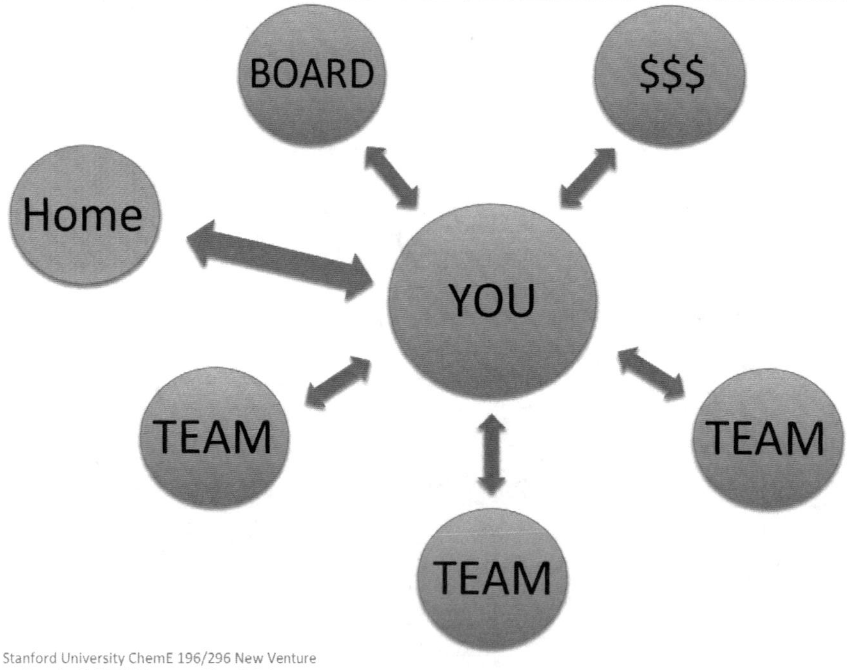

Nourish yourself

Seek mentors

Find quiet time

Meditate

Find harmony

Purpose → Actions → Values

Be grateful

Some final thoughts...

- To hire
  - Ascertain purpose
  - Use your intuition, with verification
- To keep
  - Be vulnerable but firm
- To fire
  - Decide with your mind
  - Act with your heart/soul

From the Tao Te Ching

Lao Tsu (circa 604 – 531 BCE)

From the Tao Te Ching....

In dwelling, be close to the land.
In meditation, go deep in the heart.
In dealing with others, be gentle and kind.
In speech, be true.
In ruling, be just.
In business, be competent.

In action, watch the timing.

Let's continue the conversation

www.ricardolevy.com

wng2wng1@mac.com

A Leader's Ongoing Journey –
Letters to a Young Entrepreneur
Succeeding in Business Without Losing at Life

435942 From an Investors Perspective

Abstract:

Investment leaders need deliverables that include financial metrics and risk models that reflect the possible realities of the investment. The ability to reduce risk is paramount to the successful skill set of the leadership team. The financial math distills decision-making comparative parameters to other investments. Many early pitch teams fail to communicate the salient points that drive the investment. Making money is the primary reason for all business. Business plan elements must include developing the market and the presentation of credible data that shows the current progress and the projected milestones. The multiple tranches of a chemical process or product investment adds complexity. Early investors are balanced with later money and value must continually be increased. Different firms have different time horizons and different investment metrics. The considerations and diligence for a chemical engineering investment is different and more detailed than the ones that make the news.

435943 Investment Metrics - Sorting the Wheat from Chaff

Abstract:

Today a lot is coming over the transom as entrepreneurs are flooding the world with business plans and ideas for start-ups. Investors, as well as potential partners and employees need a rapid method to identify those businesses worth a further look and to succinctly identify areas for growth for those businesses not yet ready for investment. This challenge is complicated by the wide variety of businesses being created that operate with different business models and plans for multiple industries.

A simple, rapid score sheet methodology will be presented that quantifies the way investors think. This will be presented with examples and appropriate weighting of factors so that readiness can be quantified and compared to other startups competing for investment, talent and opportunities. Common business sense and judgment will be merged with quantification. Examples and evaluations will be given for the dimensions that are scored with the methodology

436073 The Data Sets: Eos, Klrs, and Measurement Metrics

Abstract:

The critical element to the technical piece of the investment is the data that supports the claims. Can the product be demonstrated to be reproducible on the systems that are proposed ? The capital investment for chemical projects are large and models must be validated against actual data from bench and pilot systems. Experimental Orders (EOs) and Key Learning Reports (KLRs) are a simple and organized way to define the runs that are required to produce the data that supports the claims being made for the investment. Capital and operating costs must be justified to the financial investors to deliver the returns expected.

436076 Intellectual Property Development

Abstract:

Having a great idea that can evolve into a business is a fragile thing until there is a critical mass in cash flow to sustain the business growth. Intellectual property can help protect your innovations while your business develops and can also help you secure market position and generate revenue. But building a strong IP portfolio to protect your ideas requires thorough documentation and a coordinated strategy. Recent changes to the USPTO procedures for challenging patents make it even more imperative to carefully select what ideas to protect and what type of intellectual property will best safeguard your ideas. Patents allow you to exclude others from practicing your claimed invention for up to 20 years, but require detailed disclosure and thus teach everyone what you are doing. A trade secret is an alternative to a patent that can potentially protect your innovations indefinitely as long as they remain secret, but cannot protect you from reverse engineering or independent invention by a competitor. Developing a strong IP portfolio that effectively protects your core business requires careful planning and strategic decision-making.

436065 Financial Matters

Abstract:

Investors need to continually improve the overall margin of the pools of money that they collect as they sell their shares in a particular fund. Entrepreneurs must present a series of commonly applied metrics and measures so that the financial independence of the investment can be compared to similar type ventures. The range of investment metric targets are fluid and are interdependent upon the goals of the fund, the tolerance for risk and the measureable combination of the various metrics of the fund. Most of the numbers presented prior to revenue and profit are estimates as to what will occur once operational systems are producing for the market. The tolerance for the chaos that can de-rail the steady state march to profitability must be considered.

435945 The Exit Strategy - Built to Sell or Built to Last

Abstract:

So when do you exceed your capacity to continue to contribute to the enterprise ? Are you a builder or an organizer ? Businesses go through different cycles of need. The leadership team that can launch the business might not have the skills to sustain and execute the growth and operationalization phase. Recognizing when you have contributed you best to the phase of the project needed for success and then realizing how to exit with the rewards is critical to closing out and moving on the next exciting opportunity.

426856 Assessment of the CO_2 Capture Potential from Irreplaceable Industrial Sources

Abstract:

Assessment of the CO₂ Capture Potential from Irreplaceable Industrial Sources

Peter C. Psarras, Jennifer Wilcox, Energy Resources Engineering, Stanford University, Stanford, CA

Abstract

In 2013, CO₂ emissions from all industrial processes totaled 163.0 MMT, equivalent to the annual CO₂ emissions of roughly 35 million automobiles. This figure excludes indirect emissions associated with electricity usage. The heaviest emitters (coal power plants, natural gas combined cycle plants, etc) continue to receive the majority of attention and funding in terms of carbon capture projects; however, these major sources may also benefit from less CO₂-intensive alternatives. Unfortunately, many industrial processes fall into a category by which there are no alternative routes to product available. For example, steel and cement production both involve processes that directly emit CO₂ as a by-product (via the oxidation of metallurgical coke and conversion of calcium carbonate to lime, respectively). As there materials constitute the irreplaceable fabric of industrialization, CO₂ emissions from these, as from other irreplaceable industrial processes, are projected to increase unabated. With capture technology in place, these emissions can be diverted instead to viable CO₂ reuse and sequestration opportunities, such as oil refining, enhanced oil recovery, food processing, metal treatment, and fertilizer production.

To assess the capture potential of irreplaceable industry, it will be necessary to geo-reference these sources alongside all current and potential future CO₂ users (sinks), with the goal of making economically sound linkages between source and similar-sized usage markets. This will entail a cost analysis of on-site capture plus additional transport costs (freight versus pipeline, hazmat fees, etc.). Geographic information systems (GIS) mapping will assist in defining the most cost-effective mechanisms for CO₂ delivery. As these costs are inventoried, the financial incentive gap necessary to compel the targeted source-sink pairings to move forward is calculated. This effort will develop a current economic assessment of moving irreplaceable industry toward carbon-neutrality. Though these industries represent a small portion (*ca.* 3%) of total CO₂ emissions, their permanence requires immediate attention. As these low-hanging fruits are tackled, this study may serve as a model for assessing carbon-neutrality in other sectors.

433370 Strategic Planning and Market Analysis; How the Natural Gas Futures Market Will Need Solar Power to Compliment Texas State Power Grid Demands

Abstract:

Strategic Planning and Market Analysis; How the Natural Gas Futures Market will need Solar Power to Compliment Texas State Power Grid Demands

Zohreh Ravaghi Ardebili*[1], Max Kolbe[2], John Sutherland[3], James Keeter[4] *[1] zohreh@keetersutherland.com, [2] maxkolbe74@gmail.com, [3] jms@keetersutherland.com, [4] jsk@keetersutherland.com

Abstract

Given the lack of market stability in the oil and gas sector, the concept of applying solar energy as a secondary or tertiary and stable power supply to state and national grids is what should be driving investors towards this market.

According to the definition, a primary energy source is an energy source that can be consumed directly or converted into some other useful power, such as electricity. Roughly a third of the natural gas consumed in the United States goes into power plants for the production of electricity. Although in the current atmosphere of low gas prices due to high supply from shale and low demand the market may seem prime for electricity production, however the horizon for natural gas market will greatly shift to undersupply and high prices. Studies show that shale reserves are short lived and the recent construction of several ethylene cracker plants in the US alone and the multiple LNG export plants in construction will have an untold and very high demand for natural gas. Natural gas prices will rise significantly and given the latest reduction in drilling the demand will not be satisfied and therefore prices are expected to rise significantly.

The state of Texas is the highest power consuming state in the US, in 2013 had a total of 367, 546 Million kW-hrs over 40% greater than the state of California which has by far the most advanced solar powered grid in the US. As a matter of primary power, sources of electricity generation should be considered and optimized. Electrical power production via natural gas fired plants may in the near future not produce cheap electricity as the chemical and export demand of it may exceed the low price capacity point and shift demand significantly upwards. Natural gas is the building block of many industrial processes and chemicals for the manufacturing industry. Consuming it for power is not strategically making use of the resource and not providing a great return on investment, whereas using natural gas to sustain a cheap and abundant supply of basic industrial chemicals is in fact building a more competitive economic structure for the US.

The goal in power production and market stability should be to minimize the use of natural gas as power producing fuel and maximize it to produce the necessary chemicals used for the manufacturing sector which in turn creates far more employment. In its place, power should be produced from solar energy; Texas is one of the largest producers and consumers of power given that it has a large industrial sector that has a high demand for power. Texas is also one of the best regions in the US for high solar radiation levels, therefore it is a logical Texas be the state where solar power be given all the required financial and technical attention to strengthen the supply of abundant and cheap solar power production.

This paper will attempt to first carry out a market analysis in the consumption and production of energy and its future trends, then followed by an application study of solar power to these trends specifically assessing both photovoltaic and concentrated solar power production system solutions. Given that Texas is one of the greatest power producing and consuming states, we will present our study for the state of Texas as a case study on how solar power can be feasibly considered as a stable power producing source while allowing the oil and gas industry to prosper and maximize its potential and strengthen the US industrial sector.

406355 IRI2038 Futures Study: A Journey into the Future of R&D Management

Abstract:

As part of the Industrial Research Institute's 75th Anniversary Celebration, it commissioned the IRI2038 Futures Study, a futures initiative designed to answer the following question: *How will possible future developments and events impact the art and science of research and technology management over the next 25 years?*

After examining hundreds of trends, weak signals, and implications, the Futures Study was able to create four plausible, yet provocative scenarios about the future of R&D and innovation management:

- Africa Leapfrogs Developed Countries
- Everything's In Beta
- Three Roads to Innovation
- Death of Distance vs. Megacities

These scenarios are not predictions of the future. They are tools to enable practitioners to better prepare for the real future as it unveils itself. The more a particular theme appears across all four scenarios, the more likely it is that it will come to fruition. Key themes to watch include: the death of IP, augmented humans, the era of women, and the end of the time zone problem.

Presentation of this study includes the process of doing a foresights study of this size, a description of each of the scenarios, and key trends that R&D practitioners need to consider. Breakout sessions can include a backcasting exercise, a great way to bring the scenarios to life and get people thinking about how they can actually apply the themes touched on in the study.

435396 Disaster Risk Hydrometeorological : Case Xalapa, Veracruz

Abstract:

This document is intended to disseminate knowledge about the dangers and identification of disaster risks in the country derived from hydrometeorological phenomena, particularly flooding, taking as a study case Xalapa, Veracruz. The types of previous disasters have as origin a natural phenomenon, so they are often called natural disasters, although in development and consequences has a lot to see the action of man.
Natural and anthropogenic disasters contributed to elevate levels of poverty and inequality. The disaster reduction is understood as the set of actions for prevention, mitigation and preparedness. Human actions on the environment aggravate and increase the negative impact of disasters.

This project is intended in the integration of the existing knowledge about the occurrence of phenomena and the use of geo-information techniques in the zoning of threats from natural events and transfer to institutions involved in decision-making in the management and regional development; as well as appropriate regulations and legislation.

For 50 years the occupation of land use for infrastructure has spread throughout the basin and is expected to expand more in the near future, which is projected in the maps presented in this regard. The risk of flooding and landslides has increased due to urban growth within the basin. The infiltration capacity of the precipitated water has declined and has increased the surface runoff, because with a precipitation in the low area are unable to evacuate rainwater, this precipitation data in past decades were greater for the occurrence of the stagnation, however the rainy season remain normal (except for hurricanes where the intensities are much higher).

422012 Integrated Process Analysis and Design for Sustainable Local Resource Management

Abstract:

Integrated process analysis and design for sustainable local resource management

Elias Martinez-Hernandez

Aidong Yang

Department of Engineering Science, University of Oxford, UK

Melissa Leung Pah-Hang

Matthew Leach

Centre for Environmental Strategy, University of Surrey, UK

ABSTRACT

Integration between facilities in a particular locality has been explored between industries as a way to optimize economic and physical resources, leading to lower waste generation, shared responsibility with the environment and the society. The most widely known concepts come from industrial ecology such as industrial parks and industrial symbiosis. These initiatives have primarily focused on the industrial or technological processes and how they can form symbiotic relationships and recycle loops. However, the processes in two key parts of a wider system are often dismissed: ecosystems (as ultimate resource provisioners) and consumption by the society. When considered at the local scale, these components have the potential to (and ideally they should) be included into the analysis and decision making for an efficient and sustainable management of resources. This paper reports a systems engineering approach for the integration of local processes or activities, as a way to use locally available resources (biomass, solar irradiation, water, etc.) in a more efficient manner to meet local population needs. System dynamics modelling is used to determine ecosystem constraints which are then fed into the models for water and energy integration using superstructure optimization. The tools and methods are combined into a framework to facilitate the identification of opportunities for exchange of resources between local ecosystems, production and consumption components and to analyze the performance of the resulting localised production system in respect to the level of satisfaction of the needs for food, energy and/or water.

The application of the framework will be shown in a case study for the integration of the food-water-energy nexus in the Whitehill and Bordon eco-town in the UK. The limits for the available biomass and water resources are set from a system dynamics analysis of processes in local heathlands and aquifer as the ecosystem components. Nutrient exchange occurs in the food production processes, which is in turn integrated with the water and energy processes using a superstructure optimization approach. The optimization uses the concept of cumulative exergy as unifying quantity for accounting resource consumption and it is used as the objective function to minimize. The case study shows how the integration between technological and ecological processes can be achieved through the recycling of nutrients and water within and across the various system components to keep the ecosystem functions that provide the fresh resources for the relevant production processes. The proposed framework can potentially provide guidance to practitioners and decision makers for the sustainable management of local resources.

437114 Taking the Whole Family Overseas: Experiences and Tips for Survival

Abstract:

Moving a large family from the U.S. to Australia brought challenges and taught lessons that may be valuable for others considering international assignments. Issues like education, housing, food, pets, finding friends for children, transportation, healthcare, and other issues need to be considered and managed. A variety of personal stories and their lessons will be shared.

421116 Managing Innovation with Partners and Teams from India

Abstract:

India is a land with rich opportunities for chemical engineers, but proper preparation and knowledge is needed for Westerners to work most effectively in India or to maximize success with teams in India. Vital tips to prepare for working in India or collaborating with teams in India will be shared, including guidance on how to maximize innovation success.

421089 Innovation with Chinese Teams: What Western Companies Need to Know for Success

Abstract:

In the West, there is a common but incorrect stereoptype about the lack of innovation and innovation skills in China. In reality, Chinese engineers, scientists, and entrepreneurs can innovate and invent remarkably well, though there are some cultural barriers that can reinforce old stereotypes and hinder productivity when Western managers work with Chinese teams. Here we will share experiences in innovation and provide critical guidance to help Western companies and managers tap the innovation power of Chinese partners and associates.

CHEMICAL ENGINEERING & THE CHALLENGE OF INDUSTRIALIZATION OF AFRICA.

DR. N. JOHN ERINNE, FNSChE, FNSE

PAST PRESIDENT, NIGERIAN SOCIETY OF CHEMICAL ENGINEERS (NSChE)

&

MANAGING CONSULTANT, CHEX & ASSOCIATES, LAGOS, NIGERIA

PRESENTED AT THE

2015 AIChE ANNUAL MEETING

SALT LAKE CITY, UTAH, USA

NOV. 11, 2015

ABSTRACT

Most of the African continent is today faced with the dilemma of massive unemployment and excessive dependence on imports, even in the face of abundant primary materials. This paper explores the prospects for deploying Chemical Engineering skills as a tool for spearheading the industrialization of the continent and resolving this dilemma. It also proposes arrangements and structures for strengthening the profession and practice of Chemical Engineering in Africa, in order to position it to play the anticipated role and achieve its potentials.

Keywords: Africa, Chemical Engineering, industrialization.

1. AFRICA: HISTORICAL PERSPECTIVES

The continent of Africa, with a very sizeable population of over one billion, consists presently of 54 independent states and two partially recognized/unrecognized states **(APP. I)**[1]. As is well known, most of the continent came under colonial rule by the great powers of Europe (U.K, France, Belgium, Portugal, Spain, Netherlands, etc.) by the turn of the last century. In the aftermath of the Second World War however the clamour for decolonization of the continent began to gain momentum, eventually culminating in the wave of independence of African State in the 1960s.

The post-independence history of the continent has nevertheless been characterized mostly by instability, sometimes bordening on chaos and anarchy, as the various states grappled with the grave challenges of self-governance and nation building, with very limited back-ground experience. This naturally also had a negative impact on the economic development of the continent. After 40-50 years however, many of the states are now settling into stable, democratic structures, more amendable to economic and social development.

It is pertinent nevertheless to observe that there are notable exceptions to this historical pattern in the continent. Whereas countries such as Egypt and Ethiopia never came under colonial rule at any stage, colonization of South Africa developed through a completely different trajectory.

2. AFRICA: RESOURCES & ECONOMY

The continent of Africa is indeed very richly endowed with abundance of natural resources, in form of:

- Petroleum (oil & gas)
- Solid Minerals
- Agricultural Products

i. Petroleum

Africa has been producing oil & gas actively since the late 1950s and early 1960s. Most of the activities were in North Africa (Algeria, Libya and Egypt) and on the West Coast (Nigeria, Cameroun, Equitorial Guinea, Gabon, Republic of Congo and Angola) for several decades. Lately, several other countries have also joined the train and are actively producing oil & gas. These include: Sudan, South Sudan, Ghana, Ivory Coast, Chad, Namibia, Tunisia, DR. Congo, Mauritania, etc. Others have as well recently struck oil and/or gas and are in the process of developing the reserves for production (Uganda, Kenya, Tanzania, Mozambique, Morocco, Sierra Leone, etc.).

Libya, with proven reserves of 47 billion barrels of crude oil, has the largest reserve in Africa, while Nigeria is the leading producer, with output of about 2.5 million barrels per day **(APP. II)**[2].

With a total proven reserves of crude oil of 165 billion barrels in 2010, Africa is estimated to have nearly 12% of the total world reserves of 1,383 billion barrels[3]. Besides, Africa accounts for about 10 million barrels per day (bpd) of crude oil production, representing about 11% of total world production of 90 million bpd[3].

Similarly, of the total world proven reserves of natural gas of 6,609 trillion cubit feet (tcf) in 2010 Africa had 520 tcf, a very significant 8% share[3].

In particular, Nigeria and Algeria, with reserves of about 175tcf and 154tcf respectively, have the largest reserves **(APP.III)**[4] and rank 8th and 9th globally.

Unfortunately however the total oil refining capacity of Africa is only about 3.1 million bpd **(APP. IV)**[5], representing barely 30% of its crude oil output and a paltry 3.6% of global capacity[6].

ii. <u>Solid Minerals</u>

The range of endowment of the continent in solid minerals consists of:

- Coal

- Other Non-Metallic Minerals: limestone, kaolin, talc, diamond, phosphate, bentonite, barytes, etc.
- Metallic Minerals: gold, copper, iron ore, bauxite, tin, zinc, platinum, cobalt, etc.

For example, South Africa with proven reserves of about 48,000 million tonnes of coal and annual production of 270 million tonnes ranks 6th in the world, **(APP. V)**[7].

iii. **Agricultural Products**

Ivory Cost is the world leader in the production of cocoa, with output of 1.4 million tonnes in 2012, followed by Ghana with 870,000 tonnes **(APP.VI)**[8]. The effect is that Africa accounts for a very impressive 70% of total world cocoa production of 4,000 million tonnes per annum **(APP.VII)**[8].

However, again, Africa lags in respect of processing capacity, with only a 16.5% share of world capacity **(APP.VIII)**[8].

On the other hand, for coffee, though less impressive, Africa still contributes about 14% of estimated world output of 7.4 million tonnes. Ethiopia, Ivory Coast and Uganda rank 5th, 11th and 12th respectively among the producing nations[9].

These further illustrate the huge endowment and potential in Africa in respect of primary materials production and processing. Indeed, in addition to cocoa and coffee, the continent is an important producer of a wide range of other agricultural crops, such as: cotton, tea, sugarcane, rubber, palm, groundnuts, cassava, sisal, maize, millet, sorghum, rice, tobacco, etc.

iv. **Water Resources**

Furthermore, Africa is very well watered by several rivers which provide water for agriculture and for power. Notable among these are such rivers as the Nile, Niger, Senegal, Congo, Orange, Limpopo and Zambezi.

v. **Colonial Economy**

In the colonial era, the economy of Africa was essentially set up to obtain these primary materials required to sustain the industrial societies of the European powers. By this very uneven division of responsibility, Africa was reduced to merely growing primary crops or extracting vital minerals while industrial activities were domiciled in Europe. Coal obtained from Africa assisted in sustaining electric power generation in Europe, just like cocoa, coffee and tea grown in African sustained the beverage and confectionary industry. Similarly, African cotton was the mainstay of the textile industry while palm produce and groundnuts provided vegetable oil used in soap making and for food.

The gap between resource availability and processing capacity for a number of commodities highlighted above clearly underlines this phenomenon.

Notably too, the finished products from these industries ultimately found their way back to the African market, mostly at prices disproportionate to the African primary materials prices.

vi. **Post-Colonial Economy**

This economic pattern has persisted, long after the decolonization of the continent. Even with the advent of petroleum, the pattern has remained unaltered. The result is that today, Africa exports crude oil to the world market and imports refined petroleum products. What an absurdity!!

If the colonial powers were however to blame for the initial aberration, they cannot be held accountable for the sustenance of the same pattern of trade several decades after independence. We in Africa must therefore accept responsibility substantially for the perpetuation of poverty inherent in this lopsided pattern of trade.

vii **Poverty**

To buttress the issue of poverty in Africa, the relative GDP per capita of Africa and the other developing regions of the world are presented in **APP.IX**[10]. Clearly, Africa is at the bottom of the ladder, despite its rich endowment with natural resources, as illustrated earlier. Even the current average annual growth rate of about 5% in the continent, as impressive as it is, is not expected to close this gap in the foreseeable future.

3. IMPERATIVES OF CHANGE

In the light of the foregoing, it becomes imperative in this 21st century that Africa must change this economic order, for her survival. With increasing number of African countries finding oil & gas in their territories, Africa's crude oil must be refined for both local consumption and export to the world. Similarly, before we export natural gas overseas, we must use our gas to generate electricity to support local industry, as well as feedstock for petrochemical industries making vital chemicals for both African consumption and export to world markets. Indeed, Africa can readily become a hub for such vital industries as petrochemicals, textiles/garments, fertilizers, aluminum products, chocolates/beverages, rubber products, etc. it is only by this that the critical developmental issues of massive unemployment, value addition to primary materials, creation of wealth and reduction of poverty can begin to be effectively addressed. It is indeed by this that the relatively high growth rates currently prevailing in the continent can be sustained and have true meaning in the lives of the ordinary people of Africa.

Attainment of such lofty goals however cannot be easy. It will certainly require massive mobilization and deployment of vital input and precursors necessary for gainful exploitation of the abundant primary resources. These inputs include:

- Finance/Funding
- Technology/Know-How
- Human Capital

4. PUBLIC INFRASTRUCTURE

It is also pertinent to note that envisaged industrialization of the continent cannot indeed be reality without the simultaneous development of public infrastructure to support it. This will be in such vital areas as:

- Electric Power
- Fuels & Energy Supply
- Water Supply
- Telecommunications
- Information Technology
- Ports/Harbours
- Airports
- Railways
- Roads/Highways

- Sewage/Waste Management
- Environmental Protection Systems
- etc.

Whereas impressive progress may have been made lately across the continent in such areas as telecommunications and information technology, critical gaps are identifiable in most other areas, between the colonial/post-colonial infrastructure available and the requirements of modern infrastructure to support industrialization. Massive deployment of resources is therefore envisaged in this area, along with the focus on industrialization.

5. ENTER CHEMICAL ENGINEERING

Chemical Engineering, as we know full well, may be defined as that aspect of endeavor concerned with the transformation of primary materials into finished products required for consumption. By this, Chemical Engineers are essentially involved in development, design, construction, operation, maintenance and optimization of processes for transformation of materials from the raw stage to finished products. Indeed, it may be pertinent to note that such processes do not always necessarily entail chemical reaction or chemical changes. Often they are essentially physical processes. Separation of air to obtain oxygen and nitrogen or extraction of vegetable oil from oil seeds readily come to mind in this regard. There are of course numerous others. This however clearly illustrates why Chemical Engineers are often referred interchangeably as Process Engineers. It also underscores the versatility of the profession.

In view of this, it is evident that Process Engineers have the unique training and background to drive the much desired economic emancipation and industrial development of the African continent. I am yet to see or know any other profession that is as uniquely positioned in this regard.

Though Chemical Engineering may not be able to provide answers to the question of financing, it certainly can drive technology and know-how, in most respects. It can also provide the vital technical manpower required in most of the sectors.

Without doubt, Process Engineers are best positioned to drive the refining of crude oil into petroleum products or processing of natural gas for use in power generation or even transforming natural gas into petrochemical products and agricultural fertilizers. A few more strategic examples will suffice to fully buttress the point:

- Production of sugar from sugarcane.
- Production of cement from limestone
- Smelting of aluminum from bauxite
- Extraction of vegetable oils from oil seeds and refining of same.

The list may as well be indeed endless.

Besides, with their training and skills Process Engineers are also eminently positioned to assist African nations structure optimum resource development and utilization strategies and plans for a very broad range of primary materials.

In addition, Chemical Engineers are also equipped to contribute in varying degrees to the development of the public infrastructure required to underpin these resource improvement initiatives, as highlighted earlier. In particular, Chemical Engineers have a strong role to play in such areas as fuels & energy supply, thermal power generation, alternative energy development, water treatment, urban waste management and environmental protection systems.

6. CHEMICAL ENGINEERING IN AFRICA

Internationally, Chemical Engineering is a late comer as a distinct Engineering discipline, relative to the more traditional branches of Engineering: Civil, Mechanical and Electrical. It's existence dates back to only a little over a century ago when it emerged and was recognized as a distinct professional discipline in USA and UK.

The pattern is also somewhat reflected in Africa. This is underscored by the earlier emphasis on basic public infrastructure development in the colonial and post-colonial periods in the continent. The focus was on hydro-power, transportation (harbors/airports/railways/roads) and communications infrastructure, etc, just barely adequate to support the colonial economy which was centered on export of primary commodities, with little or no interest in processing and local value addition. This clearly culminated in a subdued role for Chemical Engineering in the development process in most parts of the continent.

However with the compelling dynamics of change in the continent clearly pointing in the direction of local value addition and processing of primary materials, a major boost is expected in the role of Process Engineers in the continent in the coming years. This is also further buttressed by the increasing role of Process Engineers in modern public infrastructure development.

The highest concentration of Chemical Engineers in Africa presently are in South Africa, Egypt and Nigeria, expectedly coinciding with the three largest economies of the continent. A distinct and vibrant Chemical Engineering society exists in each of these countries. In the rest of the continent, Chemical Engineers do not appear to have the 'critical mass' for independent or semi-independent existence yet and are therefore understandably subsumed in the wider Engineering societies.

It is expected that the changes and developments anticipated in the economies of African nations should help to further energize the Chemical Engineering formations across the continent, both in countries where Chemical Engineering associations exist and those where they do not exist, presently. In this regard, it is imperative that Chemical Engineering is properly positioned to play the expected role in the industrial development of the continent.

7. RECOMMENDATIONS

A few recommendations are proferred, which may assist in realization of these lofty expectations outlined above.

i. **Development Strategy**

African nations are enjoined to more than ever before design their development strategies to leverage on the abundant primary materials to grow manufacturing industry, in order to combat the menace of poverty and unemployment in the continent.

ii. **Education/Training**

African nations are also encouraged to establish Chemical Engineering schools in their Universities, tailored to their needs for resource improvement and industrial development. Special attention should be paid to the curricula, to ensure that they strongly reflect the specific needs of their economies and not merely designed along the lines of the broad, classical Chemical Engineering curriculum. Besides, such schools must be well resourced with faculty and facilities, to ensure quality training. Indeed, quality is perhaps considered more important than quantity with regards to training of future Chemical Engineers in the continent, for fairly obvious reasons.

iii. **Professional Development**

In as much as current thinking clearly favors private sector lead in industrial development, it is necessary for African countries to have the mechanism in place to ensure that young Chemical Engineers and other related technical professionals have adequate opportunity for professional development and career growth. This should be considered to be an important part of governmental regulatory function which is a vital ingredient for success of the free enterprise system. It is indeed essentially the means to ensure that these professionals mature in their chosen fields in due course and are able to play the roles expected of them.

iv. **Industry Support**

The emerging industrial sector in Africa is also urged to strongly support governmental efforts in education, training and professional development of Chemical Engineers and allied technical cadres, as indeed an exercise in enlightened self-interest.

v. **African Chemical Engineers Forum**

There is compelling need for Chemical Engineers in the continent to organize themselves into a forum or association which will serve as a platform for regular exchange of ideas and experience. Such a body will also facilitate the much desired robust contribution of Chemical Engineers to the industrial development and economic integration of Africa. The existing strong Chemical Engineering bodies in South Africa, Egypt and Nigeria readily provide the nucleus and platform for such an organization. The responsibility therefore clearly devolves on them to get together and lead the way in this regard. The Nigerian Chemical Engineers in NSChE are willing and ready. Are our brothers and sisters in SAIChE and ESChE (Egyptian Society of Chemical Engineering) also similarly willing and ready? This is a call to service for mother Africa. May we indeed not be found wanting!

8. **CONCLUSION**

As Africa transits to the stage of rapid industrial development using its vast raw materials as feedstock, a central role is foreseen for Chemical Engineers as key drivers of the process. It is therefore critical that all stakeholders take the necessary steps to ensure that this role is fulfilled. In particular,

special attention to the education, training and professional development of Chemical Engineers is called for, from African governments. Professional Chemical Engineers in the continent are also specially called upon to organize themselves more effectively, to position them to play the expected role.

REFERENCES

1. CIA Directorate of Intelligence, The world fact book: "Africa", 2012.
2. US Energy Information Administration/International Energy Statistics, Oil & Gas Journal, 2011.
3. Christina Katsouris, Energy Intelligence, Africa's Oil & Gas Potential, 2012.
4. OPEC, Natural Gas Proven Reserves by Country, 2013
5. Energy Information Administration (EIA), Form EIA-820 "Annual Refinery Report", 2014
6. British Petroleum, BP Statistical Review of World Energy, 2013.
7. British Petroleum, BP Statistical Review of World Energy, 2008
8. ICCO, USDA, Reuters, 'LMC Report', 2012
9. USDA, Hawaii Farm Facts, 2012.
10. World Economic Forum, The Global Competitiveness Report, 2013.

APP. I

AFRICAN NATIONS & POPULATIONS

S/NO	COUNTRY	POPULATION	%OF POPULATION
1	Algeria	33,333.267	3.25
2	Angola	16,941,000	1.65
3	Benin	8,439,000	0.82
4	Botswana	1,839,833	0.18
5	Burkina Faso	13,228,000	1.29
6	Burundi	7,548,000	0.74
7	Cameroon	17,795,000	1.73
8	Cape Verde	420,979	0.04
9	Central African Republic	4,216,666	0.41
10	Chad	10,146,000	0.99
11	Comoros	798,000	0.08
12	Democratic Republic of Congo	75,507,308	7.36
13	Djibouti	906,000	0.09
14	Egypt	84,550,000	8.24
15	Equatorial Guinea	504,000	0.05
16	Eritrea	5,880,000	0.57
17	Ethiopia	85,237,338	8.31
18	Gabon	1,384,000	0.13
19	Gambia	1,517,000	0.15
20	Ghana	23,000,000	2.24
21	Guinea	10,057,975	0.98
22	Guinea-Bissau	1,586,000	0.15
23	Ivory Coast	17,654,843	1.72
24	Kenya	34,707,817	3.38
25	Lesotho	2,067,000	0.20
26	Liberia	4,128,572	0.40
27	Libya	6,036,914	0.59
28	Madagascar	18,606,000	1.81
29	Malawi	12,884,000	1.26
30	Mali	13,518,000	1.32

31	Mauritania	3,069,000	0.30
32	Mauritius	1,219,220	0.12
33	Morocco	35,757,175	3.49
34	Mozambique	20,366,795	1.99
35	Namibia	2,031,000	0.20
36	Niger	13,957,000	1.36
37	Nigeria	174,507,539	17.01
38	Republic of Congo	4,012,809	0.39
39	Rwanda	7,600,000	0.74
40	Sao Tome & Principe	183,176	0.02
41	Senegal	11,658,000	1.14
42	Seychelles	80,654	0.01
43	Sierra Leone	6,144,562	0.60
44	Somalia	9,832,017	0.96
45	South Africa	52,981,991	5.17
46	South Sudan	8,260,490	0.81
47	Sudan	36,787,012	3.59
48	Swaziland	1,032,000	0.10
49	Tanzania	44,929,002	4.38
50	Togo	7,154,237	0.70
51	Tunisia	10,102,000	0.98
52	Uganda	27,616,000	2.69
53	Zambia	14,668,000	1.43
54	Zimbabwe	13,010,000	1.27
		1,021,398,191	99.58

Partially Recognized/
Unrecognized States

55	Somaliland	4,000,000	0.39
56	Sahrawi Republic	266,000	0.03
		1,025,664,191	100.

APP. II

AFRICA PROVEN RESERVES OF CRUDE OIL & PRODUCTION - 2010

	COUNTRY	RESERVES (billion barrels)	PRODUCTION ('000 barrels per day)
1.	LIBYA	47	1,750
2.	NIGERIA	37	2,520
3.	ANGOLA	13.5	1,940
4.	ALGERIA	13.4	1,700
5.	SUDAN	6.4	112
6.	EGYPT	3.7	680
7.	GABON	2.2	242
8.	EQUATORIAL GUINEA	1.7	350
9.	CONGO REPUBLIC	1.8	275
10.	DR CONGO	0.8	115
11.	CHAD	1.2	17
12.	MAURITANIA	1.5	16
13.	CAMEROON	0.7	77
14.	OTHERS	2.5	180
	TOTAL	165	9,974

• includes South Sudan

APP. III

TOP TEN WORLD PROVEN RESERVES OF NATURAL GAS - 2010

RANK	COUNTRY	RESERVES (TCF)
1	RUSSIA	1,623
2	IRAN	1,120
3	QATAR	837
4	TURKMENISTAN	583
5	USA	315
6	SAUDI ARABIA	273
7	VENEZUELA	184
8	NIGERIA	175
9	ALGERIA	150
10	AUSTRALIA	128

APP. IV

AFRICA OIL REFINING CAPACITY 2010

RANK	COUNTRY	CAPACITY ('000
1	EGYPT	680
2	SOUTH AFRICA	490
3	ALGERIA	450
4	NIGERIA	445
5	LIBYA	340
6	MOROCCO	155
7	IVORY COAST	70
8	KENYA	65
9	SUDAN	50
10	ANGOLA	45
11	CAMEROON	45
12	GHANA	45
13	TUNISIA	35
14	SENEGAL	25
15	ZAMBIA	25
16	GABON	20
17	CONGO REPUBLIC	20
18	CHAD	20
19	NIGER	20
20	ERITERIA	18
21	TANZANIA	17
22	MADAGASCAR	12
23	MAURITANIA	10
		3,102

APP. V

LEADING WORLD RESERVE & PRODUCTION OF COAL 2007

PROVEN RECOVERABLE RESERVES (million tonnes)			PRODUCTION (million tonnes per annum)		
RANK	COUNTRY	RESERVES	RANK	COUNTRY	OUTPUT
1	USA	246,643	1	CHINA	6,396
2	RUSSIA	157,010	2	USA	2,537
3	CHINA	114,500	3	INDIA	478
4	INDIA	92,445	4	AUSTALIA	394
5	AUSTRALIA	78,500	5	RUSSIA	314
6	SOUTH AFRICA	48,750	6	SOUTH AFRICA	269
7	UKRAINE	34,153	7	GERMANY	202
8	KAZAKHSTAN	31,279	8	INDONESIA	175
9	POLAND	14,000	9	POLAND	146
10	BRAZIL	10,113	10	KAZAKHSTAN	94

APP. VI

WORLD PRODUCTION OF COCOA 2012

REGION	COUNTRY	PRODUCTION ('000 tonnes)
AFRICA	Cameroon	220
	Ghana	870
	Ivory Coast	1,400
	Nigeria	230
	Other Africa	81
TOTAL AFRICA		**2,801**
ASIA/OCEANIA	Indonesia	500
	Malaysia	18
	Other Asia	-
TOTAL ASIA/OCEANIA		**623**
AMERICAS	Brazil	185
	Ecuador	170
	Other Latin America	208
TOTAL AMERICAS		563
TOTAL WORLD		**3,987**

APP. VII

REGIONAL SHARES OF COCOA PROCESSING (GRINDING) 2012

REGION	OUTPUT ('000 tonnes)	% SHARE
EUROPE	1,554	39.1
AMERICAS	865	21.8
ASIA & OCEANIA	897	22.6
AFRICA	657	16.5
	3,973	100.0

APP. VIII

REGIONAL SHARES OF COCOA PROCESSING - 2012

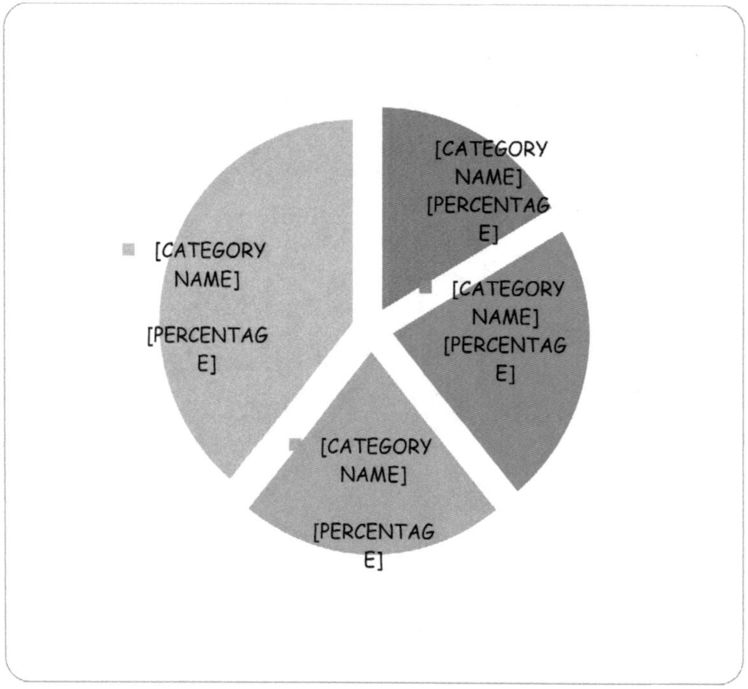

APP. IX

GDP's OF WORLD DEVELOPING REGIONS 2012

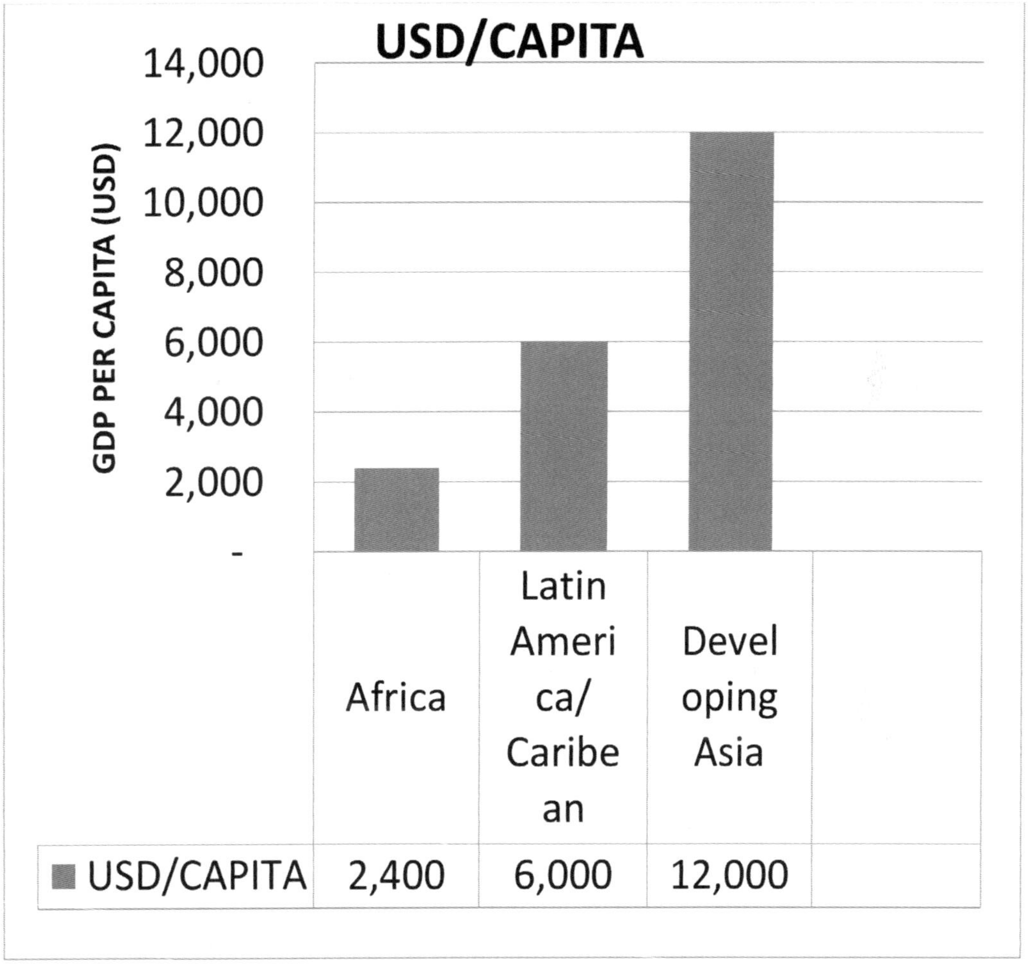

Figure 1USD/CAPITA

AUTHOR INDEX